D1063258

Farming

Gavin Sprott

NATIONAL MUSEUMS OF SCOTLAND

Published by the National Museums of Scotland
Chambers Street, Edinburgh EH1 1JF

ISBN 0948636 69 6

© Trustees of the National Museums of Scotland 1995

British Library Cataloguing in Publication Data

A catalogue record for this book is available from the
British Library

Series editor Iseabail Macleod
Picture research Susan Irvine
Designed and produced by the Publications Office of
the National Museums of Scotland
Printed by Ritchie of Edinburgh

Acknowledgements

llustrations: Front cover, 26: National Galleries of Scotland. 15: Kinross-shire
Antiquarian Society. 20: J Wilson. 22: Mrs Grace Smith. 28: Mrs Macdonald. 37:
Barbara Robertson. 42: Mrs H D McLeish. ii(top right): Dundee Art Galleries
and Museums. 49: Mrs Isobel Smith. 56: St Andrews University Library. 64:
Biggar Museum Trust. 68: Mrs Anne Gordon. 77: Mrs Doctor. 82: Ness
Historical Society.

Thanks to NMS Photography.

Illustrations captioned SEA are from the Scottish Ethnological Archive in the
National Museums of Scotland.

Front cover: *Detail from* A Hind's Daughter *by James
Guthrie, 1883.*

Back cover: *A pair of Clydesdales in show harness, water-
colour painted in Banffshire, 1910, by the horseman who
looked after them.*

CONTENTS

1 Old bones

There are still places where you can stand and imagine what it was once like: in the distance cattle browsing knee-deep among the coarse grass, the scent of bog myrtle, the fiery trill of the laiverock somewhere above, and nearby the quiet blether as people work away at the peats. Moments when time seems to stand still. Or there are others when the plough skins the stone lid off a grave that has lain undisturbed for thousands of years, and there lies the crumpled skeleton of what was once a living human being, but so remote in time that she or he might have come from another planet. These figures from either the living past or chance symbols of death in the present have one thing in common – they are and were farmers. Wandering hunters first colonized what is now Scotland after the last ice-age, perhaps 8,000-10,000 years ago. They were followed by the New Stone Age farmers, some of whose descendants will be working the land today with tractor-mounted ploughs and combine harvesters.

The farming that evolved from these early times relied heavily on the use of natural resources that lay to hand, and used them for fuel, building, and above all the natural herbage was used for grazing. Over the centuries the emphases changed. The first settlers made rapid inroads into the forest. Their descendants were overtaken by the change to a colder and wetter climate which started the growth of the peat that was to become such a vital fuel. But overall they devised a way for a thin scatter of population to exist over a wide area. The cultivation was on the best free-draining land. It was cropped continuously, as we still crop our

Milking a native Highland cow, Outer Hebrides, early twentieth century. SEA

*The crofting township of Uig, Lewis, just before or after
World War II.* Alasdair Alpin MacGregor, SEA

vegetable gardens to this day. Its fertility was maintained or even
enhanced by using the muck the animals generated from the
pasture and putting it on the arable. Thus, the livestock was the
means by which the good of the poorer land could be concentrated and used to best effect on the better land. By the eighteenth
century the improvers looked back with horror on this old
farming. They were seeing something that had broken down
under the weight of population growth, and where the natural
resources were being consumed more quickly than they could
regenerate. Yet there was a time when it was the most intelligent
way to do things.

With different ways of working went different attitudes.
Animals were not the crop of butcher-meat of modern farming.
They were kept for their products, of milk, wool, and energy for
cultivation and transport. Only when they became a wasting
asset, when they could not be kept through the winter, were they

killed, preserved and eaten. Livestock might be eaten when they died of disease, or the odd sheep killed for a celebration. Even then, nothing was wasted. Hide, bones, sinews, intestines, fat, horn – all had their uses, down to the bladder which could be given to children for a football.

Later generations would scorn the old-style farmers for not improving their livestock through breeding. But if there was one single character prized above others, it was hardiness, and related to that, resistance to disease. The hungry months of winter and early spring were the most ruthless breeder of all.

Pasture was a common resource, reflected in the general name for it – *commonty*. In the Lowlands it might also be called *muir* or moor. By far the greater part of the countryside was grazing. It was not lush meadow land, but *naitur gerss* or natural 'wild' grass and much else besides: sedges, rushes, heather and so on, the mixture that to this day gives mutton from the hill its rich flavour. People were limited in the amount of livestock they could have. That was worked out in proportion to the arable land that they rented, and was their *souming*.

The pasture was used carefully, with the animals actively herded on one part or another according to the season. In the spring the livestock would go out beyond the *heid dyke* to the commonty. Some cattle might be brought back in along the *loaning* or main thoroughfare morning and evening for milking, or so that they could drop their muck on land yet to be ploughed. The furthest pastures were the spring or summer *shielings* which made best use of the shorter growing season in the hills.

The cultivated land at the heart of any settlement went by various names, including *croft* land, or *muckit* land, reflecting the treatment it got. At the centre was the *infield*, always cropped, and then the *outfield*, periodically cropped until the return did not repay the effort, then left for several years to recover its fertility. The outfield could serve other important purposes, as a quarry for *fail* or sods of turf widely used for building, and as a reserve of winter pasture.

7

Drainage was one of the greatest difficulties. Lack of skill and technology kept cultivation off much of the richest ground in the valley bottoms and carselands. The ground was ploughed into *rigs* or ridges, so that water could run off into the spaces in between. These spaces could develop into *bauks* where grass, whins or broom might grow. The latter appeared as weeds to later generations, but the old-style farmers had good uses for them.

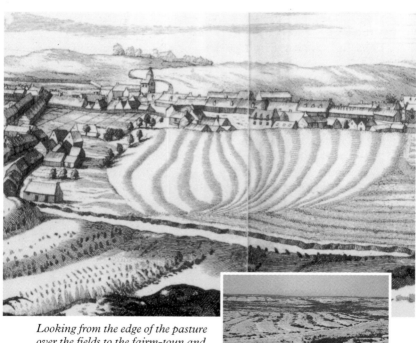

Looking from the edge of the pasture over the fields to the fairm-toun and the burgh of Arbroath beyond, late seventeenth century. From John Slezer, Theatrum Scotiae, *1693.*

Old pre-improvement rigs show up in the snow near West Calder.

Whins were a reserve fuel, often used by the poorest people. *Bruim* made a good thatch. Other prolific weeds such as *thrissel* were carefully harvested and made a good feed for horses.

The range of crops was small, but had an internal balance. *Corn* or oats, and *bear*, a hardy form of barley, were the main grain crops. Some wheat was also grown on the richer ground right up the east coast as far north as Orkney. *Pease* and beans were very important, both for their nourishment and for the nitrogen that they fixed in the soil. The *pease strae* was a valuable winter feed for the milking cows.

In fact, the diet was more varied than meets the eye. Besides dairy products, there were the eggs of domestic poultry and both the eggs and meat of the big range of wildfowl that inhabited the then extensive mires and moors. The medieval abbeys stimulated the planting of orchards, providing a wide range of apples and pears, some strains of which survive in obscurity to this day. The old kirk's demand for fine candle wax encouraged the keeping of bees, and thus of honey. The good of the bear crop was realized not just in bear-meal but in ale, a common part of most meals. And what of the tip given by King James IV to the woman who brought a present of strawberries for him and the queen?

An interesting facet of modern Scots is that it retains meanings long since lost in standard English. An example is the word *toun*. That has two meanings, one the same as standard English, and the other, a farming settlement. The Gaelic *baile* also has this twin meaning.

The *toun*, in effect a village, was at the core of the old farming. This survives in a modified form in parts of the Crofting Counties, where the *sràid* or main road forms the backbone of the settlement.

In the old touns, these holdings of arable consisted of a number of rigs, with one family perhaps having a parcel here and a parcel there. So although these rigs had specific tenants, they were still managed in a strongly communal way. The farm was usually rented from the laird by several families in a joint tenancy. To ensure that everybody had a fair chance of the best and the worst land, the rigs could be periodically reallocated by the drawing of lots. This continued in parts of the Highlands and Islands until the break-up of the old system by the Clearances and by crofting agriculture. In contrast, in the Lowlands habit increasingly prevailed, the same ground remaining with the same family from one season to the next. Overall this pattern of inter-mixed arable holdings was known as *runrig* or *rundale*.

The practical work was dominated by the pooling of labour and resources. Ploughing was very labour-intensive. In the Lowlands, cattle were the main draught animals. One person would guide the plough, another help control it, the *gadsmen* at the side urging the animals forward and another person at the front of the team leading it on. Others would come after the plough with spades to sort the *fur* or furrow. Obstinate clods that would defy any harrow would have to be bashed down with a *mell* or big wooden mallet. To achieve all this, people had to pool their animals. In the west of Scotland and in the Highlands and Islands, horses were also used, in the latter areas sometimes with several yoked side by side.

This communal emphasis pervaded everything, winning fuel, harvesting and so on. It also applied to the running of things. The laird's wishes would be expressed either by *himself* or his depute, the *baron baillie*, presiding over his baron court. The officials of the court were drawn from the principal tenants, and one of the main concerns was the use of communal assets, in particular timber, fuel and pasture. It was here that people who had sneaked more than their *souming* of livestock onto the commonty would face the reckoning, not merely of the laird's displeasure, but also of their aggrieved fellow tenants.

Communal though much activity was, it also involved great self reliance. Far from the simple life, existence was a treadmill of problems which had to be solved locally or not at all. Illness, disease in man or beast, mental instability, the dangers attending childbirth, these all had to be coped with by the knowledge to be found within what was often a small and quite isolated community. People not only had to know about the properties of medicinal plants, but where to find them, and at what season. For those things which obviously defeated local ingenuity, they fell back on rituals, which, if they made no difference to the outcome perhaps made it more bearable. For instance, on an outbreak of murrain or foot-and-mouth disease, all the fires in a district might be extinguished, fire kindled afresh from *needfire* or rubbing sticks together, and all the cattle driven through the resulting bonfire to *sain* or cleanse them.

The simplicity of the technology can be deceptive to hindsight in various ways. The process of thatching a roof may be basic, but skill in choosing the materials and applying them can make all the difference between a misery of leaks and contented warmth. Although the houses differed from region to region, there was a

Simple but effective technology. Cold water is poured over the hot boulder, shattering it. Lewis, 1928. SEA

fairly general pattern of a long narrow building which included dwelling and byre at least. The materials – an amalgam of field stone and turf, timber, brush-wood and various thatches – were all gathered locally. The buildings thus had a high organic content, and were in fact integrated with the organic cycle of farming, and were machines for reprocessing other materials into that system. For instance the peat on the fire would not only provide heat, but its soot lodged in the thatch, enriching it when it was recycled into compost. The fire's heat 'cooked' the rain-soaked thatch, breaking down the fibres so that when the old thatch was stripped off it would compost well. The peat ash was added to the litter in the byre, adding valuable potash to what

Roping down new thatch in Northbay, Barra, 1936. Coir yarn has replaced ropes of twisted heather. Margaret Fay Shaw Campbell, SEA

Peat barrow and peat-cutting tools at the Scottish Agricultural Museum.

would be spread as manure. More stone and earth tents than houses in the modern sense, these buildings had a logic that made good sense to people at the time.

When Robert Burns wrote *The Cottar's Saturday Night*, he first quoted a fragment of Gray's *Elegy*. Gray had mused about the hidden potential that had never been realized in the obscure lives of these country people. Burns celebrated the landless labourers of Ayrshire not for what they might have been, but for what they were. Although in Burns's day things were changing, with opportunities opening up for the *lad o pairts*, even then the blunt fact was that nearly all the people of whom he was talking were going to stay the way they were, in poverty. It was accepted as the natural order of things, and that fixity probably made for a kind of dour feudal familiarity between laird and people.

In the old country society, social mobility was static or downwards, so that gentry and peasantry could be of the same kindred and know it. The only chance of change for the better, a remote chance, lay in emigration to a burgh, and advancement, perhaps over several generations, to become a merchant, and then if sufficiently successful, a return to the country as a laird. For most people, hope for a better life lay in the hereafter. Their main concern was to survive, and keep what they had.

Keeping what they had was not always easy. To hang a sheep stealer or cattle-raider seems harsh, yet it made sense. To steal a family's livestock was to cripple its livelihood. Over Scotland as a whole, if people were not actively engaged in farming, there was little else for them to do. If a family suffered the disaster of a broken tenancy, they might get places as *cottarmen* or *acremen* – sub-tenants with a small patch of land, often just a *yaird*, to work as their own, and for this they would work for the tenant proper. People who did not get a place as sub-tenants were in danger of falling out of the system. The respectable destitute might be licensed to beg. Others joined the ranks of the *sturdy beggars* who roamed the countryside, an endemic social problem and often a menace to the settled population. Thus it was vital not just to hang on to livestock, but to any tie to the land itself.

Here is one vivid instance of the value of land to people. In 1567 Robert Mure, the Laird of Caldwell, sold the lands of Wester Kittochside in Lanarkshire to the eight tenants. One of them was John Reid, who became a *portioner* of one sixth of the land. Ten years later, Caldwell regretted the sale, and tried to get the land back. When Reid refused to part with it, the enraged laird harried the Reids over twenty years, culminating in two raids in which Caldwell's three sons and their followers rampaged through Reid's house terrorizing his wife and children and stabbing the beds in search of John Reid. Discovering that the guidman had escaped, they set fire to the house, and made off with whatever they fancied, including plough horses, cattle and seed corn. The Reids were reduced to begging from door to door, but John also went to Edinburgh to the king's *Secret Counsal*, or privy council, and complained. Caldwell found himself imprisoned in Edinburgh Castle at his own expense until he had made full recompense to the Reids. The Reids were left to enjoy their holding as *bonnet lairds* or small owner occupiers, and in fact enlarged it with the compensation money.

There must have been many instances when justice did not prevail. Compared to her neighbours, Scotland was not a violent country. In the centuries since Bannockburn, the number of

Cutting peat at Portmoak Moss, Kinross-shire, around 1900. This shows the whole process of cutting and drying peat. SEA

really bloody conflicts was very small. It was a country used to peace, but it was a rough kind of peace. As much as the cattle and sheep and corn on which their livelihood rested, the people had to be hardy and tough.

2 A mania for improvement.

At one time famine was an ordeal everyone expected to face perhaps at least once in a lifetime. In the late summer of 1695 a curious and continuous haar set in on the East Coast. It was the harbinger of 'the Seven Ill Years' of bad harvest and famine. No one knows how many people perished in these years, but in many districts it was at least a third of the population. In Sir Robert Sibbald's famous phrase, the living were wearied with burying the dead. Some put the disaster down to the wrath of God. Others sought the causes elsewhere.

In pre-industrial times population growth was slow, but nevertheless steady. There was thus a constant incentive to produce more grain, and this could only be done by ploughing up more of the permanent grazing. The greater the arable, the more draught animals needed to plough it, but less grazing on which to sustain them. The real bind would come in winter, when there was little or no natural growth to sustain the livestock, and not much in the way of stored winter feed. With more mouths to feed, both human and animal, the old system was thrown out of kilter and began to break down.

The task facing the early improvers was finding alternatives to relying on the self-generating natural resources, and in particular creating adequate winter feed for the livestock. The answer lay in deliberately cultivating grass, turnips and potatoes as field crops. To conservatives, this seemed a rash leap into the unknown. Where could it be done? To reduce the area already under crop to create another as yet uncertain one must have seemed too risky. So often the first improvement started on that general-purpose area, the outfield. But it could not stop there, because the crops had to be rotated, and now that these included the sown fodder, within five or more years, the new crops invaded all of the old arable and more. The *commonty* or permanent grazing also had to be ploughed up and become part of the new scheme.

Thus, the whole balance was changed. Much more ground was cultivated, but with far less frequency. The cycle started with a grain crop. Then might come a turnip crop, followed by another grain grop, this time undersown with grass seed that grew in the wake of the harvested

Model of a neep barra *or turnip sower at the Scottish Agricultural Museum. These late-eighteenth and early-nineteenth-century models were often part of the design process for new machines to deal with new crops.*

grain. For the following two years this was harvested as hay, then directly grazed as meadow for another three years. Then the time had come for ploughing up what had become the *lye* or lea and sowing it with a grain crop again. In this way no more than a third of the ground was under the plough at one time. As the rotations extended, this proportion could sink to below a quarter.

It was difficult to change in a self-contained farm-by-farm way. Where a commonty served several farming townships, if one changed, they were all pushed towards change. The Parliament of Scotland provided for this by laying down a simple procedure. If the majority of *heritors* or landowners in a parish agreed, the commonty could be split up between them. This in itself was significant, because although it took more than a century for this process to be completed over the Lowlands, it showed that the underlying mechanics of the new system were understood early on.

Initial progress was slow, limited to areas where heritors with the means and enthusiasm to change were the majority. The first improvements became effective in East and Midlothian in the 1720s, and appeared in isolated spots up the East Coast. But once there was sufficiently widespread capital and understanding, the balance was tipped towards very rapid change. In the Lowlands this point was reached about 1760.

The state of the ground was a formidable obstacle. The old strains of corn and barley were habituated to the ill-drained acid soil, but not so the likes of turnips nor the mixtures of clover, rye-grass, cocksfoot and timothy that made up the new sown crops.

That was when the landscape we recognize today began to be created. Because it has the mellowness of age, it is easy to imagine that it all evolved gradually. In fact, the changes that came with improvement were drastic, and they involved a truly massive effort of physical labour. The first thing to be tackled was the abundance of stones in the ground. Over the years these had been rolled down off the rigs to accumulate in the spaces between, trapping silt and blocking what drainage there was. They could be gathered into *consumption dykes* several feet broad. The *big yirdfest* or earthbound stones that defied spade and pinch-bar were shattered by gunpowder. Trenching was an intensive digging over of the ground, and was often done by travelling squads of men from the Highlands. This brought up more stones and cleared the ground of weeds. The whins and broom that had an economic use in the old-style farming were dragged from the ground by yoking a horse to a *dug*, a wooden beam with an iron jaw to catch the base of the stem.

For the time being the rigs remained, because the main drainage in the fields was still by surface run-off. However, the rigs were substantially remade with shovel and barrow, their high crowns scaled down and their crooked plan reformed into neat straight rows, between fifteen and eighteen feet broad, the cast of a sower. In the beginning, enthusiasm was sometimes ahead of experience, and good topsoil got buried.

Even though drainage was thus improved, it remained the improvers' biggest battle for over half a century. A first attempt was made at sub-soil drains, either stone box drains, or *rummel syvers* or stone-filled soak-aways, and at the field boundaries more ambitious attempts were made to run the water off into burns and rivers through open drains. Larger schemes of drainage could change whole districts in a spectacular manner, as with Blair Drummond Moss in south Perthshire, or the lowering of Loch Leven in Fife.

The use of lime was known before improving times, but now it came into its own. The effect was to neutralize the acid in the soil, and thus make the nutrients available to the crop. The limestone was broken down into small stones and burnt. The simplest although least efficient way was to do this in the fields. By the mid-century multiple *kills* or kilns were being built on an industrial scale, situated at points where there was ready land and sea transport for stone, coal and the finished product, as at Limekilns on the Forth in west Fife. An alternative to lime was marle, which was like a dirty and hard but crumbly chalk, sometimes found in pans where it had accumulated millions of years before.

As the old grazings were broken up and integrated with the arable, how were the animals to be separated from the growing crops? The answer lay in *enclosure*, building permanent barriers between one field and the next. Here was a whole new trade, hedging and dyking. The common hedge was the *may* or hawthorn. To make it stockproof, the vertical boughs were layed by half-cutting them and bending them down to form a dense network. Dykes took many forms, becoming part of regional character. Simplest of all was the use of children to herd the beasts, the commonest expedient in the early days of improvement. It could take surprisingly long for the process of enclosure to be completed, often into the 1860s.

With enclosure came timber wind-breaks. Even where there were no plantations, field boundaries were seen as a place for

*Burning lime at Gladhouse, Midlothian, 1935. The kiln
had been refurbished to provide lime as part of much-needed
help to agriculture at that time.* SEA

another crop: useful timber. Ash was the most versatile timber,
light for its strength, easily worked and tough. Depending on
area, oak, beech and *plane* or sycamore also became popular field-
boundary trees.

As cultivation spread, the former pockets of arable eventually
linked up in the patchwork of enclosed fields that we now recognize. In some areas, such as Buchan and Easter Ross and the
more fertile glens of the Highlands and Islands, improvement did
not happen on any scale until the nineteenth century. Other parts
of the *Gaidhealtachd* would follow a different route into the *via
dolorosa* of the Clearances and beyond.

Changes in the way the land was used changed the society that
used it. In the seventeenth century, the structure of old-style
farming had in some places undergone a modest evolution that

helped this change. More people worked the same infield over the years instead of it being reallocated at the start of each season. Single rather than shared tenancies increased, as did the size of some holdings.

The more enlightened improving laird was seeking not just a tenant and rent, but in effect a business partnership. The new tenant had to have sufficient capital to stock and equip the place adequately. The new *tack* or lease was a careful document, guarding against the repetition of old ills, such as using *fail*, and with that the precious topsoil, for building purposes, or over-cropping the ground. In return, there would be the promise of long and sometimes heritable tenure, so that the farmer could work to hand on something worthwhile to his family. Capital invested in the permanent structure of the farm in draining and dyking was allowed for in adjustment of the rent, or low interest loans.

The kind of man the laird or his factor sought was literate and open to the new ideas, such as Andrew Wight, who farmed near Ormiston in East Lothian. Sufficiently impressed by his intelligence and skill, the managers of the estates confiscated after the Jacobite Rising of 1745 engaged him to do reports on their properties. So useful were these that Wight went on to produce his *Present State of Husbandry*, front-line reports giving a vivid account of improvements in various parts of Scotland as they happened.

By the European standards of the time, a significant number of the Scottish nobility and gentry worked to earn their privileged position. A man such as Sir John Sinclair of Ulbster was tireless in gathering and disseminating practical information, the lifeblood of any intelligent advance. There was an atmosphere of discussion and education. The Society of Improvers, which ran in Edinburgh from 1723 to 1745, included 300 of the most influential people in Scotland. From 1784 the Highland Society acted rather as a development agency, prodding, advising and stimulating an exchange of ideas. The Chair of Agriculture founded at Edinburgh University in 1790 was the first in Britain and the second in Europe.

Bondagers *or women outworkers, probably East Lothian between the world wars. The two men are the* wumman gaffers. SEA

There were also those lairds who went through the motions of improvement without understanding its substance. One of Burns's bitterest complaints was against the Ayrshire lairds who half improved their land, and then demanded a rent appropriate to an East Coast farm in full production, as in the case of Mount Oliphant.

The change was even more drastic for the bulk of the population. The world of the numerous small tenants and their subtenants who consumed most of what they produced would vanish. Most families no longer sought ground on which to raise a crop and keep stock, but a fee or job with a wage. Thus labour relations in the countryside were becoming industrialized before the Industrial Revolution started in the towns.

The fee was a six or twelve month contract, and provided the basic stability on which the new system thrived. Over the years, regional variations became marked. In the South-East the whole family was involved. It was difficult for the *hind* or ploughman to get a *single fee* or employment on his own account. His wife and daughters were *bondagers* or women outworkers whose labour was vital from the planting of potatoes and singling turnips in the spring to harvesting them in the late autumn, and above all during the *hairst* or main grain harvest. On the hill farms of the Borders, a shepherd and his son often worked for a *dooble fee*. From east-central areas to the north, the six month fee for a single man became the backbone of the labour force. On the smaller farms of Buchan, it became common to hire one man who slept in the *chaalmer* and got his food in the farm kitchen. On the bigger farms, and particularly so in Angus, several men shared the *bothy*, in effect a two-roomed barrack in the steadings.

Feein Setturday *at Arbroath. The stalls were selling* fairings – *confectionery, snacks etc.* SEA

Customs that were remnants of the old-style farming survived. In Ayrshire the *bower* or cattleman might keep some cows on his own account, and likewise the Border shepherd might have his own *pack* of sheep. There were various perquisites that could go with a fee: *favour milk*, a supply of potatoes, grazing for a cow, the keep for a pig.

Some of these could be bargained for at the *feeing markets* or *hiring fairs* that were held before the May or November *terms* or break points in the contract. The *fairm sairvants* developed an ethos of life and work which is still vivid in living memory. In the eighteenth century it was something entirely new.

3 Speed the plough.

In 1710 a Scots joiner found himself on a most unusual journey for someone in his situation in life – an expenses-paid trip to the Netherlands to study the construction of pot-barley mills. For his sponsor, Andrew Fletcher of Saltoun in East Lothian, it would be a good investment. James Meikle came back with the information, and the Fletcher family established a profitable monopoly of pot-barley milling. But Meikle had also noticed something else – winnowing machines. He made one when he got back, in effect a big box with paddles inside turned by a crank handle. The revolving paddles created a blast of wind, and as the grain fell in front of it, the chaff and lighter grains were blown to one side and the heavier grain dropped straight down. But instead of regarding it as a success, intelligent people denounced it as producing the *Deil's wind*. Who should presume to create extra wind when divine providence had not seen fit to send it?

To suggest to a nation of Calvinists who believed in predestination that you could boldly design your way round fate was astonishing. For many years to come winnowing would depend on a good draught across the threshing floor, and if that were not sufficient, the grain was taken to an exposed spot and winnowed in the wind. Such places are still remembered in place names such as *Shilling Law*. Only in the 1730s did *fanners* or *winnisters* as they were known begin to find some acceptance. This underlines a simple fact: the basic act of improvement relied less on technology than on the way the land and resources were used. The technology developed to multiply the advantages offered by improvement.

James Small was born in Berwickshire in 1741, and there he served his time as both wright and smith. He used this combination of skills to the limit. He also had something more than skill – an original and observant mind. He spent some time in Rotherham, in Yorkshire, working as a plough-wright. There he saw a local plough which had been improved by Dutch designs, and that set him thinking. Small returned to Berwickshire, and about 1763 he was working on his own design. Instead of the *reist* or mouldboard (that part of the plough which turns over the furrow-slice) being flat, he experimented with a slightly screwed shape. Using boards of soft timber, he charted the patterns of wear, and made adjustments until the pattern was even. He also tested the resistance to draught using spring balances. Once he had arrived at the best shape, he sent his patterns to the Carron Ironworks near Falkirk, which had opened just four years before, and back came mouldboards cast in the curved shape that tractor ploughs still have to this day. The improved swing plough was born.

Small's plough was an enormous success, and from the later 1760s it was spreading into the main arable areas of South-East and Central Scotland. The design was widely copied and adapted, helped by his own publication on the subject. Realizing the lift this would give improved agriculture, Small forbore to patent it. As Sir John Sinclair remarked, Small 'had such a propensity to be useful that he laid personal interest aside too much', for he died hard-up and worn out, at the age of 52.

Now the efforts to improve the state of the ground had progressed sufficiently for a lot of land to be much easier to plough – lighter, better drained, cleaner, and cleared of big plough-break

Turning the plough at the heid-land *or end of the park or field. The team is now a* pair o horse. James Howe. SEA

ing boulders. Here the new plough came into its own. The other big change was the switch from oxen to horses. Horses were more expensive to buy and feed, but they were faster and more intelligent. One crucial advantage was that a horse's hoof would hold a shoe much better than the cattle beast's soft cloven hoof. The horses could as readily go on the new metalled roads as work in the fields.

The horse team was much shorter and more manoeuvrable than a yoke of oxen, another advantage that went with working within the new parks or enclosed fields, where there was now a dyke or hedge to limit turning space at the *heidland*. At first, four horses yoked in two pairs one in front of the other was the replacement for the oxen. This still required one person to lead the horses and another to guide the plough, the way that Robert Burns was used to working. Then as the big and powerful new Clydesdale breed spread towards the end of the century, the team was reduced to a *pair o horse* worked by one man, the common standard except on some crofts and smallholdings, where one horse had to do.

In this simple fact alone lay a whole labour revolution. Farming improvement led to cultivation over a much bigger area, but it was done with less manpower. The simple equation was one man to a pair of horse, one pair to 50 or so acres on average arable land. Nor was it just numbers. Before, ploughing was something everybody bore a hand in. Now there was one man out there, skilled in the ways of his horses, deftly turning over the ground until the regular *seed furs* of crumbly tilth resembled a well-ordered garden. The old word for a ploughman in the South-East was *hynd*. Now another word appeared throughout the Lowlands, not as a surname, but as a job description, the *horseman*. Working with the new Clydesdale horses and the improved ploughs had become a skill in its own right.

This is reflected in the birth of the ploughing match in the late eighteenth century. Initially these competitions were got up by the newly-founded agricultural societies and encouraged by interested lairds and the bigger tenant farmers. They soon took on a life of their own, becoming popular annual events. There were prizes for the best work, the best turned out pair o horse and the handsomest ploughman. Smiths turned out special match ploughs with long boards and special *crankit* coulters and bridles or hitches that could be finely adjusted. The preparation of harness started not days but weeks ahead, burnishing the leather and metalwork, and perfecting the various decorations. The champion ploughman was king.

Another reflection of this new skill was the Horseman's Society. These were not formal organizations, but a loose brotherhood, with a similarity to Masonic organizations. The *horseman's oath* bound the initiate to secrecy, and imparted the *horseman's word*, which when whispered in the animal's ear would produce a wonderful obedience. Horsemen's societies provided solidarity among the men, and were a source of companionship and fun. There could also be a less attractive side, a hard time for the young and vulnerable initiates, and cruelty towards horses to induce the appearance of obedience. The societies were most evident in the bothy areas on the East Coast.

Agriculture differs from most other industries in one important respect. The production takes place over the face of the land. Small's plough was the first step to making that production as efficient as it could be. But unlike coal and iron production, or manufacturing, it could not be concentrated, and would prove difficult to mechanize. Indeed, achieving full mechanization was going to take two centuries. However, processing the product could be concentrated, and grinding the grain had already been simply mechanized for centuries.

James Meikle's fanner and pot-barley mill had shown what ingenuity could do, but working with mostly wooden parts, there

Ploughing match at Fearnan, Loch Tay, 1930s or '40s. The horses are turned out in well-burnished show harness. SEA

was a limit. From 1759 the Carron Ironworks opened up many other possibilities besides the mouldboards of Small's plough. The old timber *lantern* or cage gearings could be replaced with cast iron *nicketie wheels* and bevels. Mills could be re-designed on a much grander scale. Better water-wheel design gave more power and enabled mills to be built on sites that provided inadequate water before. Corn-drying *kills* or kilns became part of the local mill rather than a place in the farm steadings, made possible by the perforated cast-iron *kill-plates* on which the grain was spread to dry. By the late eighteenth century the old compulsion of being *thirlt* or tied to the estate's mill was fast breaking down, as the miller's trade became part of the new market farming.

As the new farming burgeoned, new bottlenecks appeared, and one was threshing the grain. Various devices were tried that imitated the action of the hand flail. None worked and some exploded in a shower of splinters. The answer lay not in imitation but genuine invention, and that fell to Andrew Meikle, the son of James. His revolving drum with its rows of pegs that knocked the grain from the straw was an immediate success, and from the late 1780s it spread through the main arable areas of Scotland and beyond. The oldest surviving threshing mill in the world is preserved at the Scottish Agricultural Museum. Erected at Breck of Rendall in Orkney about 1804, it was probably prefabricated and shipped up from the Bo'ness area. The castings are crude by modern standards, but they gave good service for over a century and a half.

The skills of the millwright were a specialization of the *wricht's* or joiner's. The country joiner now came into his own. His workshop was often near the smiddy. It took both trades to make the new box carts essential to the new farming, and in the earlier days of the improved ploughs, the frames were still made in timber. Increasingly people did less specialized work for themselves and left more to the tradesman. A clear example of this is the saddler's trade. In the Lowlands, the do-it-yourself harness people devised from timber, straw and even sods of earth that went with the old

Ringing a cart wheel at Innerleithen smiddy. Two men hammer the heated iron tyre into position while a third cools it with water. SEA

slypes and tumbler-wheeled carts was by 1790 being superseded by substantial leatherwork, fit for the bigger horses and box-carts, and for yoking the new ploughs. Until that time, the saddler would have had little trade in the countryside beyond the requirements of the gentry.

It is interesting that by 1800, in the more developed districts, those directly involved in cultivating the soil were often about only half of the rural Lowland population. Many of the other half settled in villages, some old villages rebuilt and many totally new creations. Besides the wrights, smiths and others, and those involved in the expanding building trade, there was a range of occupations that were not strictly trades yet still involved hard-learned skills. These included dyking, ditching and draining. The more domestic trades included those of the *baxter* or baker, the *flesher* or butcher, the *souter* or shoemaker, and the tailor.

Some of these occupations had long been common in the burghs, but the appearance of bakers, for instance, in the countryside reflected a new fashion for the luxury of wheaten loaf, pushing out the common *breid* of oatcakes. More money and personal property also meant work for *notaries* formalizing transactions and drawing up wills, as well as the means to afford the services of a local apothecary or surgeon.

There had always been *wabsters* or weavers in the countryside, producing cloth from wool yarn spun and dyed at home, but in the late eighteenth century their numbers grew considerably because the invention of mechanical spinning produced masses of spun yarn without, until the 1840s, efficient power looms to weave it. This gap was filled by handloom weavers. Another subculture often based in the countryside, but with a character of its own, was that of the colliers or miners. Until 1799 many of the miners and their families were serfs, a kind of brutal job-creation scheme imposed in the previous century. By the end of the eighteenth century serfdom was no longer profitable to the masters, and it was abolished. The free mining communities developed in part out of the old serf communities, and in time earned their reputation of being tough, resourceful and politically aware.

Miners were often part of the rural community, working at various seasonal jobs on the land such as planting potatoes, singling turnips and harvesting, but they never lost a certain separateness. Local farmers waged war against their dogs, especially at lambing time, and miners and gamekeepers were constant enemies, sometimes with tragic consequences.

Perhaps the biggest change of all was work for landless labourers. From the 1760s the new turnpike roads generated a steady stream of work. Their construction demanded labour and the hire of horses and carts to shift spoil. Once they were built they had to be maintained, and a man with a horse and cart could set up as a carrier. The new rural industries such as quarrying, tile-making and saw-milling needed labour, and the trades that fed off them also needed labourers to work to the tradesmen.

One group that lived neither in fairm toun nor village but in their tents were the *traivellin fowk*. They were also known by the settled population as *tinkers*, and were and are vulnerable to their whims and hostility. Yet in the changing countryside they also had their place. They had particular skills of tin-smithing and basket-work, and made many useful domestic items from *pot-reenges* or heather pot scrubbers to clothes-pegs, and decorations such as paper flowers. Where the same people returned every year for regular jobs, for instance mending the *spale sculls* used when the potatoes were lifted, suspicion could be replaced with mutual respect, and sometimes kindness.

The *Big Hoose* itself was changing. The fruits of farming improvement were in many cases reinforcing fortunes made in the sugar or tobacco trades. A consequence of Scotland's position within the United Kingdom was that the political manager Henry Dundas gianed a dominant position in the East India Company. Dundas used his ample patronage to buy votes, and in Sir Walter Scott's words, the Company became the 'corn chest

Maids from the milk houss *and the* big hoose *flank the shepherd with his plaid and dog. Possibly the Borders, early twentieth century.*
SEA

for Scotland'. Half a lifetime under a foreign sun might improve an old estate or buy a new one.

Even the smallest estate was a miniature kingdom, the laird and his lady the monarchs, and the factor their prime minister. Everything was dependent on servants: fires, cleaning, laundry, cooking, gardening, transport, building and grounds maintenance. When the laird walked out with his gun, it was over ground cleared of vermin and stocked with game by keepers and underkeepers. When the lady drove out to visit, it was in transport maintained by stable-boys, grooms and coachmen. This was only the most obvious aspect of a veritable chain of servitude. Most numerous were the fairm sairvants, the men and women who were fee'd to labour the ground. A tradesman's wife might have a lassie to do odd jobs, break sticks for the fire and clear the ashes. The women served the men, and the children served both. It was a way of life that would last until World War I.

4 Newbiggin

John Younger was born in 1785, in the village of Longnewton, near Ancram in the Borders. His parents had a holding of fourteen acres in what he described as a 'village farm'. They were one of the six tenants holding between thirteen and 30 acres. He remembered the day when he was a child that the 'land doctor' came. The prescription was that all these little holdings and four others would be turned into one farm, and just one of the previous small farmers would become the sole new tenant. His father was the local shoemaker, and was allowed to stay on renting the house alone, but the ground was ploughed right up to the walls, and to rub in the sore, the new farmer came and shot the three remaining hens lest they stray onto his new tenancy. Within a few years the old village of twenty households had wasted away to almost nothing. In other cases, the change was even more abrupt. Tenants were told to quit at the term, and their houses pulled apart and the turf and thatch that came out of them spread as manure.

Neat new farmhouse and steadings at Gogar Bank,
Midlothian, 1795. The cornyaird *is filled with a bountiful*
crop, and useful timber flanks the farmer's genteel garden.
From George Robertson, Agriculture of Midlothian, 1795.

Why such drastic upheaval? The way the new farming worked
forced a reshaping of the settlement pattern. Instead of the old-
style toun or semi-collective farming village, the new bigger single
units would have what at the time were described as *centrical*
farmhouses and steadings. Often these were built on or near previously occupied sites and inherit the old names. Other new farms
might emerge grandly as New Mains or similar. Apart from the
Big Hooses, the manses and the kirks, there was virtually a clean
sweep of the old. Only in the Lothians and Orkney did a scatter of
older farm-related buildings survive this change. It was indeed
the time of *newbiggin* – new building.

The new building was mostly stone and mortar. The mortar
was of two kinds, the older clay mortar – in effect mud, sometimes mixed with sharp sand and straw – or lime mortar. Another
less common yet significant construction was using clay like mass
concrete. Many clay or clay mortar buildings still stand, but their
construction is not obvious, because they have been harled or the
stonework has been pointed with lime mortar.

New roofing materials were employed, on the East Coast red pantiles, and in Angus and East Perthshire the heavy grey stone slate quarried in the Sidlaws. The blue Scotch slate of Argyll would not be used on buildings outside the coastal towns until the railways came to the countryside. Overall, still by far the commonest roofing was thatch, but improved kinds of thatch such as reed, or the old materials used in a new way.

The availability for the first time of straight-sawn timber in quantity was the key to new roofing and more than one-storey construction. In the old-style buildings, the timber had been felled and split with wedges, and advantage taken of the natural bends to form the curved blades of the three or four *couples* or big frames that supported the roof. To tile or sark and slate or employ the improved thatches over such an uneven foundation would have been impossible. But however grand or simple the new buildings, the quarried stone and sawn timber had to be transported.

There was a road system of a kind in the old countryside before improvement, but a system of routes would describe it better. Transport needs were not just less, but very different. Cattle, sheep and the pack-horses used to transport grain and general goods did not need roads as we know them. People either walked or went on horseback. Only a great nobleman might use a carriage, and where that could go would be very restricted. There were carts, small-bodied with *tumbler* or solid timber wheels, and *slypes* or sledges, used mostly for local or internal transport. The main roads were meant to be maintained by the local populations, but in practice little happened. The main obstacles to movement were the rivers. From the Middle Ages Scotland had fine bridges at the important crossings, often built for the public good by local benefactors. The first 'modern' roads were built in the Highlands following the 1715 Rising, but their purpose was military, not economic.

All this had become totally inadequate, underlined by the consequences of enclosure. Once fields were dyked or hedged in, people had to stick to the road, and the old roads actually got worse before they got better. The Turnpike Act of 1751 signalled the start of a new transport system. This enabled the formation of local trusts, which raised the capital and charged tolls. At first local lairds were often suspicious of the new roads passing over their ground. By the 1780s they were squabbling about who should get the good of them running through their estates, as the economic benefits were by then so obvious.

The new system had to provide an alternative to the old reliance on local fuel and timber, and that meant connection with the coalfields of the Forth and the West, and with the ports. The coal would be needed not just for domestic use, but to burn lime, which also had to be transported. The rising output of the new farming had to be got to market. The new buildings required large amounts of materials, most of them bulky and heavy.

A thait- *or trace-horse assists with this box-cart laden with wool-sacks. The new turnpike road leads to the distant mills and factories. From Henry Stephens,* Book of the Farm, *1855.*

Rebuilding part of the Perth-Inverness road, about 1900.
Improvement of communications was a factor in agricultur-
al development from the eighteenth century.

The new roads were made with careful drainage, good bridges, *shod* or paved fords, and were finished with *metal* or small stones and blinded with grit to bind them into a solid surface. The routes were surveyed and designed to gradients that could be readily worked by laden carts. There are still many relics of these roads, particularly the toll houses with a wing lying into the road, where there would be a *check-bar*, only raised once the fee had been paid. This was a natural place for people to stop for refreshment, which accounts for the frequency of 'Auld Toll' as a pub name.

The two-wheeled box-cart developed as the backbone of rural transport. It was becoming common in the 1750s, and until the main spread of railways into the countryside between 1840 and

1860, it was the only way of shifting heavy goods. When the heavier Clydesdale horse spread from the 1790s, the carts got bigger, some with bodies that would *cowp* or tip independently of the shafts. Many other variations developed, including flat-bodied carts for the harvest – the *shelment cairts* of Fife, the *lang cairts* or *boat cairts* of the Lothians, and the *jenny linns* of the South-West. There were the *spring cairts* with which small merchants travelled the country roads with their goods, and the milk carts to get that product to the towns and later to the station. The two-wheeled cart became widely used outside Scotland, and was often known as a Scotch cart.

The cart was the dual product of joiner and smith. Then there was the shoeing of the horses. A horse doing a lot of road-work might have to go to the smiddy as often as once a month. Even if there was no significant wear on the shoe, it would have to be removed at least every two or three months – that is the growth in the hoof paired down and the shoe put back on.

The new roads caused a fairly general switch to horses from oxen for farm work, indicating how important transport was to the new farming.

Just as the old-style buildings had their logic, so did the new. The early improved farm steadings that survive appear small-scale to our eye, but they seemed extravagant to people at the time, in their size and permanence. They were in fact a basic part of the operating plant of the new farming, essential for handling, processing and storing the new and bigger range of crops, pro-cessing the greater output of milk, stabling the horses and housing the beasts in winter. The first improved buildings were often built by the tenant with a loan or allowance from the laird.

The permanence of the buildings marked the end of their fabric being part of the organic cycle. Another significant change was social. The new farmhouses had their backs to the steadings, were often two-storied, and could be physically separate, more so in the eastern Lowlands. The tenant farmers and the *wark fowk*, the people they employed, became more socially distinct.

Cottar folk at West Barnes, Dunbar. East Lothian farms were often like small villages. SEA

This new race of farm servants was mobile. Even if farmer and worker got on well, the contract still had to be renewed every six or twelve months. Many workers chose to move – on account of a *faa-oot* with the farmer, the offer of a shilling more elsewhere, a better equipped farm, and perhaps a better house. In the new scheme of things, the housing had to be provided by the employer, and was tied to the job.

Except for the houses that went with some occupations such as cattleman or byreman, the cottar hooses were also separate from the steadings, usually in a row. The most basic consisted of one room and a garden for vegetables. New housing soon became miserable and ramshackle, and only a marginal improvement on the old. Clay floors were still standard, but the new tiled roofs were draughty and provided poor insulation, for there were often no ceilings. Even into the twentieth century, farm servants in East

Lothian had to provide their own fire-grates or ranges, and spare bricks to fill in the spaces at the sides if the range didn't fit.

In subsequent rebuildings, farm-servant housing steadily improved – a screed of concrete over the clay, strapped and lined walls to keep the rising damp at bay, a bedroom with a timber floor, a scullery, the luxury of water piped into the house, until halfway through the twentieth century the promised land of a second bedroom, hot as well as cold water and a bathroom was reached. Of the complaints voiced by Scottish farmworkers, housing rather than wages was usually the main one. The often semi-isolated existence, and bringing up a large family in tied and ramshackle accommodation made the wives frequent champions of emigration to the villages, towns or beyond.

Privacy and personal space and convenience were limited or non-existent. Yet we know from living memory and experience that those brought up in these narrow circumstances, often had impressive personal standards of honesty, courtesy and kindliness. What Burns called 'honest poverty' was the norm.

When Lowland farmworkers moved to a new fee, one or two carts were sent to convey them and all their possessions to their house. That carried all they owned: cooking utensils, crockery, the *bine* for washing, table, chairs, kists, beds and bedding, and whatever clothes they had. To the cottars of the eighteenth century and before, that would have seemed a wealth of possession. For the poorest then there was very little furniture, eked out by seats and spaces built into the fabric of the house, and for a bed people slept together on heather or suchlike spread round the fire. This was common in the Highlands and Islands well into the nineteenth century.

It was the increasing import of timber in the eighteenth century that began to make life more comfortable, whether in a modest modernizing of the old-style houses or furnishing the new ones. That enabled people to build box-beds, virtually a little house within a house, providing privacy, increased warmth and a barrier against drips from a leaking roof. A penalty was one of the scourges of the nineteenth and earlier twentieth centuries, the contagion of the consumption or tuberculosis. Even in the Hebrides, where people could little afford to buy timber, the floating detritus of driftwood from the industrial Lowlands and beyond fetched up on the shore, and would be turned not just into beds, but kists, tables, chairs, settles and dressers of simple but often elegant proportions.

The later eighteenth century saw a growing division not only between farmer and farm servants, but between farm servants and tradesmen. Often the farm servants' children would not have the benefit of a continuous and stable education, moving as they did from one school to the next. The children were always in demand for farm work, and their parents anxious for even their pennies wages. When compulsory education loomed in the years before 1872, miles of *stab and wire* pailing were rattled up all over Scotland as a substitute for the children herding the livestock. By contrast, the children of tradesmen had all the benefits of a stable location. Tradesmen and farmers were often well read, elders of the kirk and active in local affairs.

In the Lowlands in particular, this was a world to which people were already looking back with nostalgia in the late nineteenth century, with *kailyaird* stories of simple arcadian virtue. A much more balanced work is William Alexander's *Johnny Gibb of Gushetneuk*, set in 1843, a subtle tale of goodness versus cunning and greed. Even in the monoglot Scots-speaking world of small-farm Buchan, life has always been like that. Illegitimacy was common, especially amongst the vulnerable girls in service. It was almost normal for the first child to be conceived outside marriage but to be born within it. In Scotland the legitimation of children by the subsequent marriage of their parents was a licence for a certain sexual freedom before it. But equally important, the old social controls were loosened by the upheaval and new patterns of work and settlement.

Young families and big families also imposed another kind of discipline. Sharing a small space demanded instant obedience,

The bothy at Gagie, Angus. On the left is the bothy form, *on the right a* kist *or chest. The men wear* tacketie baets *and tie their* breeks *below the knee with* nickie tams *or* wull tams. SEA

and if not, instant chastisement. Children worked in some form as soon as they could, the older looking after the younger. Nursing mothers often had to continue working, their infants brought to them in the steadings or the fields by the older children to be suckled. Closeness could soon wear into harshness.

North of the Forth, and north of the Tay in particular, for unmarried young men there was the life of the bothies. The system flourished, because the farmers got good value for money in the pride the men took in their work. It was condemned by the guardians of social morality because of the men's insolence and disregard of organized religion.

The bothy, usually part of the steadings, had a bunk-house and a living-room with a fireplace. Each man's possessions went into one or later two *kists*, the *mealer* and the *claeser*. As the May

Galloway cattle grazing on the hill in Glenesslin, Dunscore. They do not gain weight as fast as modern breeds, but they can thrive on poorer ground. Gavin Sprott

The National Trust for Scotland orchard beside Melrose Abbey. The Melrose apple is still grown locally.
Gavin Sprott

A meal break at the hairst *near Auchendinny, Midlothian, early nineteenth century, by an unknown painter. Here is life from childhood to old age.*

Plowt kirn *or milk churn,* luggie *and stool for milking,* boyne *for settlin the cream and a* skimmer *for collecting it. In the Scottish Agricultural Museum.*

Kinnordy, Strathmore, 1936, by James McIntosh Patrick. The laird walking up the hill with his keeper would fall in the coming war, and the then familiar pleuchie *with his* pair o horse *would vanish soon after.*

A consumption dyke *made to absorb field stones at Kingswells, Aberdeenshire.*

Vegetable gardens are one of the oldest forms of farming: hand cultivation, intensive manuring, constant cropping. At Blackford Hill, Edinburgh. Gavin Sprott

Creaguaineach, Loch Treig. Now a sheep farm, it was built as a shooting lodge at an old but remote settlement. Gavin Sprott

or November term approached, so did *speakin time*. Then the farmer would ask the man if he wished to *bide*. If the farmer remained silent, the contract would not be renewed, and the man would have to go to the *feeing market* and seek another place. On the other hand, if the men had fallen out with the farmer or his wife, they might all go, leaving the farmer with a *clean toun*, a bad name and a new set of men to find.

The bothies had the reputation of being *roch*. This was fostered by the boisterous pranks – the raids on the farmer's wife's hens and their eggs, the treatment of the boys who were the most recent recruits, the *coorse* songs that never got into print. Yet in certain ways the bothies simply preserved old patterns long after they had disappeared elsewhere. Often enough the men ate round the fire rather than at a table. The backbone of their diet continued to be oatmeal, although that had been displaced by the potato elsewhere. They supped their brose with a horn spoon from a wooden *caup* held in the palm of their hand. Within living memory the bothy mens' possessions would expand to include a bicycle and perhaps a fiddle or melodeon. Only when the tractor became widespread after Hitler's war did the bothies disappear, and with that a trace of domestic customs that stretched back to the times before the Agricultural Revolution.

5 High farming

At eleven o' clock on a dark autumn evening in 1828 a strange contraption was pushed out of an old out-building at Mid Leoch, a farm on the southern slopes of the Sidlaws in Angus. The machine had already been tested after a fashion on a mock-up of a crop set up inside the shed, and adjustments made. So diffident was young Patrick Bell that the machine had been constructed in almost total secrecy, which would continue until it had got a proper field trial. The horse Jock was yoked behind the machine and in the half-dark started pushing it forward into the crop of wheat. To the inventor's immense relief, the machine not only worked, it made a creditable job. A few days later, on 10 September, the first successful mechanical reaper in the world had a public demonstration at Powrie, astonishing a crowd of local lairds and farmers.

Bell had got the idea in the summer of 1826 on seeing a pair of shears stuck in a hedge where a gardener had left them during a meal break. Bell made a rough model, then over the following months built the machine, with the help of the local joiner and two local smiths, and got the castings made in the East Foundry in nearby Dundee.

Bell's reaper only had a moderate success, because it was ahead of its time. The ground was still cultivated in rigs, and machinery did not manage well on this corrugated surface. Nor was there then the ease with machinery that we now take for granted. Nevertheless, one of the machines that Patrick Bell made himself and christened locally *The Lass o' Gowrie* worked well for many seasons at his brother's farm at Inchmichael. The description and plans were published and reached a wide audience. Two Americans, McCormick and Hussey, developed the idea further, and their machines would become popular in Europe a few years later.

The story of the shy divinity student labouring on his great invention in secret is not the whole picture. Bell was cautious because he knew well that others had tried and none had achieved sustained success. Their mistake was trying to reproduce the action of the sickle. Like Andrew Meikle with the threshing machine, Bell sidestepped conventional imitation, and came up with something entirely new.

The notion of a mechanical reaper was a natural one for the times. Mechanization was seen to be making dramatic progress in mine drainage, and especially in the textile industry. The man who came nearest to producing a workable reaping machine before Bell was James Smith, using the mechanical know-how he had gained at the spinning mills on the Teith water at Deanston in south Perthshire. In 1823 he started farming there. At first he

Patrick Bell's reaper. The man steers it with a pole from behind.

Model of the first design of Bell's reaper in the Scottish Agricultural Museum. This version is for one-horse working. The weight of the machine required an adaptation for two horses.

put in deep *rummel* drains between the rigs, and ploughed them level, attempting to eliminate a factor that had hindered his reaper. It didn't work, because the hard pan of clay beneath the top-soil was preventing the water reaching the drains below. In the same year that Bell produced his reaper, Smith was experimenting with a sub-soil plough. This followed in the furrow opened by a conventional plough, with a long straight knife that could reach up to sixteen inches down into the soil. This broke up the barrier of the pan enough for the surface water to percolate down to the drains. Three years later Smith published his *Remarks on Thorough Drainage and Deep Ploughing*. His discovery was an instant success, and opened the way to working ground that had hitherto defied all attempts to drain it, including the rich carselands of alluvial soil. Existing ground was greatly improved, and now the land began to take on its present appearance of a level field surface.

The spread of thorough sub-soil drainage was made easier and cheaper by using the then recently developed *mug tiles* mass-produced in clay, which were first made in Scotland in Ayrshire in 1826. By 1850 the ploughed rigs were fast disappearing from the best land. Another fifty years on and ploughing in rigs survived only in poor upland areas.

Sub-soil drainage also raised the yields. Ground that had been wet and stiff to work could now be cultivated earlier in the year, providing a better growing season and a surer harvest. Less seed was needed because it was less likely to rot in the ground. Coupled with the savings made by Meikle's threshing mill, which greatly reduced grain loss, productivity as well as output was greatly increased. Improvement was moving into top gear.

Not that it had all been a painless progression. The near quarter-century of the French Wars had made farming prosperous in financial terms, but had provided few really good harvests, and the disastrous harvests of 1799 and 1800 produced what were remembered euphemistically as *The Dear Years*. Fortunately for the farm servants, a considerable portion of their wages was in kind, and that protected them from the high prices. Despite rotten housing, that was one reason why they worked on with apparent uncomplaining industry, to the surprise of radical visitors like William Cobbett. For the others in the countryside, it was different. Increased income did not match the cost of living. In an age when income and subsistence were finely balanced, this meant a grim time for many. There was not actual famine, yet many people would have gone for months and sometimes years doing heavy manual jobs in the shadow of constant nagging hunger and malnutrition.

Nor was it plain sailing for the farmers themselves. The protection of the Corn Laws could not save the more improvident from the penalties of overcropping and careless farming that had been possible in the war years. In the later 1820s came many bankruptcies and a mood of retrenchment. It was a more sober and wiser community of lairds and tenants that stood to take advantage of the opportunities that came.

Cults is a farm near Murthly at the western end of Strathmore. Murthly lies on the old Highland Railway. Although the railway is still running, the trains no longer stop there, nor is there a goods siding as there used to be. In fact the goods siding was probably more important than the passenger service, because at one time the various carts, potato diggers and other items of farm equipment that came down the hill from Allans implement works at Cults were dispatched from that siding. Cults has reverted to what it was before – a farm. There you can still see what looks like an extension of the farm steadings. The main difference from the outside is the row of chimneys along one side that indicate the positions of what were once forge hearths inside. It was a factory in the fields. How was that possible?

Breaking up the hard pan beneath the topsoil to let water reach the drains. An ordinary plough opens a furr *or furrow, and the sub-soiler follows. From Henry Stephens,* Book of the Farm, *1855.*

A rake of new carts leaving Allan's Works at Cults, Murthly, early twentieth century. SEA

The siding at Murthly connected Cults with Scotland, the British Empire and beyond. Equally important, the world was able to communicate with Cults and buy what Allans produced. This was possible through several developments: cheap materials, the penny post, banks, and newspapers. The combination of these and railways had within a few years created a market of vast proportions. From 1849 Scottish farmers could read the weekly *North British Agriculturalist*, and peruse both the advertisements and reports of implement trials. The Scottish banking system was then the most advanced in the world, and made the transfer of payment for materials and products simple. And the hot-blast furnaces that now made cheap iron for the Clyde shipyards and engineering works also provided it for small family firms.

Sometimes the products of these small manufacturers travelled to the other side of the world. Old ploughs turn up in New

Zealand bearing the stamp *Hally* or *Barrowman*. These were made in country smiddies at Auchterarder in Perthshire and Saline in Fife. It is said that they were put into their crates for shipment in a bed of seed corn, destined for settlers from the old country. But for most of these firms, the market was more local.

For the most part, the names speak of small towns throughout every part of the Lowlands, and by 1860 all of them were connected to the rail network. It all points to one thing – increasing mechanization. The towns of the industrial revolution were totally dependent on the improved farming to feed themselves. Now the countryside also depended on the towns not just for a market, but on the small country towns in particular for the changing technology.

New ploughs outside Hally's smiddy at the Baads, Auchterarder, 1892. The Hally family had other smiddies in Perthshire and Angus. SEA

However, a basic part of the farmer's plant was still the farm steading, and what had been grand new structures in the previous century now seemed pinched and awkward to use, and because they had often been put up by the tenants on cheap loans, the lack of quality in the building now showed. Another new wave of construction started, particularly in the South East. This time the estates put up the capital, and many of the new steadings took on the monumental quality that marked big industrial buildings in the cities. There was not just largeness of scale but a kind of restrained decoration and a high standard of workmanship. Model plans were published in professional journals, and if they are now more than a little bashed about by the passage of modern machinery and hemmed in by big post-war sheds, some of the originals can still be seen, complete with their spacious cattle courts and chimney stacks where once the steam engines that drove the threshing mills were installed.

It was a matter of pride for many landowners that their tenants were housed in substantial dwellings, reflecting a chain of civilized values. New fronts were grafted onto the old, and examination will often reveal a modest eighteenth-century house standing behind a spacious Victorian one with bay windows opening onto pleasant lawns. This did not necessarily mean a widening gulf. Into this century most of even the substantial tenants spoke the same Scots as their servants, and the bulk of their servants were literate. The difference was not yet so much cultural, as a considerable divide in wealth, expectation and confidence.

There was often a significant improvement in the farm servant accommodation. The English fashion of two-storey cottages even crept in to some places. There were houses not just for the workers. At the bottom of the cottar-house garden there might now be a small dual-purpose building, one half for the privy, and the other half the pig's *crae*, with a cart-track running behind them so they could easily be mucked out. The pig would be fed on everything from the potato-peelings to dishwater, and around November it would be killed. The small part that was not pickled in salt and saltpetre would often be the only fresh meat that people ever saw – that and perhaps poached rabbits, so many rabbits that those who can remember weary at the thought of rabbit stew.

About four miles south-west of Gorebridge in Midlothian is the Rosebery Resevoir. A bit up the bank stands a small granite memorial, and on it is the following inscription: 'HERE ON DEC^R 15TH 1883, TWO BRAVE MEN, JOHN FORTUNE AND JOHN MAC-DIARMID, MET THEIR DEATH IN THE DISCHARGE OF DUTY.' Two other men also perished, at the end of a rope. They were the poachers who were convicted of the gamekeepers' murder.

In nineteenth-century Scotland few issues could have rivalled game and poaching as a sure cause for an argument. Rabbits were a menace to farmers' crops, deer roamed and increased where once there had been sheep and cattle. The exploits of famous poachers such as Sandy Davidson and John Farquharson were romanticized in print. The small window in the back of his house at Blacklunans through which Farquharson escaped his pursuers is still pointed out to little boys. Yet managing the game supported many livelihoods: keepers, under-keepers, stalkers, ghillies, baillies, deer-watchers and so on. When it came to an issue, there was little public sympathy for those who abused these employees. The widows and children of the hanged Rosebery murderers were shunned, and had to leave the district.

There have long been laws that reserve the grandest game such as deer for the grandees, but only from the early nineteenth century was poaching seriously 'criminalized' in Scotland. The severe anti-poaching laws, such as the Night Poaching Act of 1828 are usually interpreted as protecting the lairds' interests, as indeed they did, but that was far from all. The severest penalties were against poaching in gangs after dark, in an attempt to curb violence. By the same token, proprietors setting man-traps had been banned the previous year. What had happened?

In the days when the greater part of the land was rough pasture, the range of wild fowl for the taking was huge. Even then, it was not done indiscriminately. The main means of catching birds was with nets or snares. When firearms became available to the common people in the late seventeenth century, we hear complaints against the use of *fire ingynes*, not against fowling. But with the draining of Blair Drummond Moss alone went a considerable wetland habitat. As the new well-drained parks with their abundance of stubbles and convenient hedgerows and windbreaks increased, so did the numbers of pheasants and partridges. Who had paid for the changes but the lairds, and the new game they also saw as their creation. And as the towns and cities grew, having ground to shoot over acquired a social prestige which depended on being exclusive to the gentry to maintain its value. What was still a way of life for a whole rural population could for the first time be a plaything for others whose wealth came from urban industry.

The middle third of the nineteenth century was the period of *high farming*, as it was known. The land was an investment and source of wealth, the natural place to put your fortune once you had made it. It would not last.

6 The big sheep

Where there are hills and mountains, driving cattle and sheep up to the pastures for the summer months is one of the oldest customs known to farmers. It makes the best use of the land, and saves the lower pasture for the lean months of winter and early spring. Using upland ground for only part of the year could also be a steppingstone to permanent settlement, allowing ground to be broken in gradually until there was enough to sustain people throughout the year. The summer pasture had various names, *shielings* in Scots, *àiridh* in Gaelic, and *setter* derived from the Norse *setr*. These words survive mainly in placenames such as Gala*shiels*, *Ari*nagour and Mel*setter*. Going to the shielings was one of the highlights of the year, with long days and fine weather, and the freedom of the hills for the children. This we know, because in Scotland shielings survived in Lewis within living memory.

Early morning milking at a Lewis shieling. Ayrshires have replaced Highland cows. The cow being milked nuzzles a titbit. SEA

The shieling huts were usually in pleasant sheltered places near running water. The sites can often be spotted miles away, little green patches at the base of a corrie or on a raised bank in the bend of an upland river. The green is a sign of continuing fertility, because there the cattle left their muck when they were brought in for milking. A closer look at the sites will reveal rickles of stones that mark all that remains of the stone and turf huts. They are the most poignant sign in the landscape of a life that is gone.

But ancient as they were, in the eighteenth century the use of the shielings began to change. Particularly after the curse of cattle theft was reduced after the 1745 Rising, many cattle from the shielings followed the long road leading to the Crieff Tryst, or later the Falkirk Tryst, thence to be fattened by graziers in England, and eventually reaching the market at Smithfield in

London. For the Highlands and Islands, the cattle trade afforded something like an Indian summer. The drover, who acted as the middle-man between the farmers and the English cattle-dealers, would return from the Lowlands with cash in his pocket. To this might be added the proceeds from small-time distilling, and perhaps work at the hairst in the Lowlands, and this went to pay the rent.

That rent money was needed, because the principal change that followed the '45 was that, whether they wanted to or not, the Highland chiefs became Highland lairds. The prestige of an armed following was changed for that of landed proprietor, and if they were to keep up with their Lowland counterparts, they had to find more money. To turn the shielings into cattle-raising grounds was a fateful step away from the internal balance of the

Flitting cattle to the steamer, Orkney. The floors of the boats were lined with planks to prevent the beasts staving them. SEA

old subsistence economy, and towards the series of monocultures that have been the mark of the Highlands ever since. After the cattle came the sheep, the kelp, the deer, the forestry. Only time will tell if in human terms tourism and fish-farming will be the failed miracles of the twentieth century. Only crofting, a much-changed form of the old subsistence farming, has sustained either a population or the ancient Gaelic culture.

In 1707 the Union of the Parliaments opened England to a legal pursuit of the cattle trade, and following that some Galloway lairds effectively turned small farms into cattle ranches, turning off the small tenants and building dykes to form cattle parks. In 1724 there was a violent reaction, the *Levellers* flattening one enclosure after another. But the lairds had their way. Soon a regular 30,000 black cattle were leaving South-West Scotland every year for England.

The problem for the Highlands and Islands was that they were producing an increase in people as much as cattle. Whatever other factors were involved, the coming of the potato certainly played a big part in the events that were about to unfold. The potato had been grown in gardens in Lowland Scotland long before it became a field crop in the early eighteenth century. Then in 1750 MacDonald of Clanranald forced it on his reluctant ten-antry, but soon it caught on. It had many remarkable advantages. It was simple to plant and sure to harvest. Unlike a grain crop, it did not have to be sheared, bound, stooked, won, stacked, threshed, winnowed, dried and ground. It could be lifted and eaten. There was one further crucial advantage. It was eminently suitable for cultivation in *feannagan* – in the Lowlands known as *lazy-beds*, an old use of the word to indicate that the ground between the beds was uncultivated. A strip of manure – often seaweed – was laid on the bare ground, the seed potatoes placed on top, and the ground on either side flipped over with a spade or *cas chrom* to cover the seed. These beds could be made in odd spots, often in among rocks, where it was otherwise impossible to raise a crop.

Preparing lazy beds for potatoes by laying seaweed as manure. SEA

Even more important, the combination of potatoes and lazy-bed cultivation proved an easy way to break in previously uncultivated ground. With a relentlessly expanding population, the potato must have seemed a wonderful discovery. By the late eighteenth century, it had displaced the grain crop as the staple food over much of the Western Highlands and Islands.

Another sign of the increase in population was the steady miniaturization of tools and implements. When people's holdings were so reduced by subdivision that they would no longer support a horse they took to hand cultivation. The *cas chrom* or 'crooked foot', an enhanced form of spade, was a product of the late seventeenth century which came into its own over the next hundred years. If the *cas chrom* was the mouldboard, the *crann nan gad* and the *risteal* represented the functions of sock and coulter of a plough separated into different implements, workable by less draught strength.

The widespread evictions of small tenants that overtook the Highlands from the late eighteenth century have appeared in popular tradition as a terrible thunderbolt. They reflect the horror conveyed by the eyewitness accounts of people such as Donald Macleod in his *Gloomy Memories*, the eviction parties half crazed with drink firing the roofs of the houses, smashing furniture and tipping meal-kists complete with contents down the banks, the aged and infirm carried out bewildered and choking with smoke, some to die of shock. That the detail of who was evicted and who was not has been proved to be not strictly accurate by reference to estate records, and that Macleod's picture of arcadian bliss that preceded the Clearances does not stand up to examination, cannot hide the barbarity of what was done. People thought it barbarous at the time, but as they were mostly monoglot Gaelic speakers, they did not get a hearing.

The man is turning the ground with a cas chrom *or foot plough, and the woman planting potatoes.*
SEA

Eviction in North Uist, 1895, probably for non-payment of rent. By the time photographers took any interest, the worst of the Clearances were over. SEA

The barbarity was a ham-fisted attempt to 'modernize' the Highlands. In the case of the well-known Sutherland Clearances, it was accompanied by considerable sums spent on roads, bridges and what would now be known as 'infrastructure', the capital being the profits from the Bridgewater Canal, which had come to the estate by marriage. It came as a shock at the time not just because of the barbarity, but because it was the reversal of a previous trend. The proprietors had wanted the people to stay. Apart from the rental arising from the population, there was still a crude belief that a country's strength lay in its manpower. The Seven Years War had proved the value of the Highland Regiments. The restoration in 1784 of the Forfeited Estates to the heirs of the attainted Jacobites included a vague wish to resurrect a martial ethos in the service of empire. In contrast, the people themselves had been showing an increasing inclination to emigrate. In some cases where the estates did away with the *tacksmen*, they managed to persuade the tenants to leave with them.

The uproar of the Clearances has also obscured the alternatives that were tried both before and after. In 1725 a Glasgow merchant, Daniel Campbell of Shawfield, bought Islay. He started reorganizing the farming on the lines of what was to become conventional improvement, but with the scale of the new farms kept small. Kintyre, parts of Speyside and the Perthshire Highlands, and later the Highlands bordering Easter Ross got similar treatment. Some of this ground, but by no means all of it, was of better quality. It was the agricultural depressions of the later nineteenth century that reduced the populations of these places. What the earlier changes did was to stop the headlong increase.

There had also been an insidious form of clearance taking place south of the Great Glen since the middle of the century – the letting of hill grazings. People were not brutally expelled from their farms, but by depriving them of their summer pasture, the old system was effectively decapitated. Onto the old shielings came the *Linton* or Blackface sheep. They were not so hardy, and their wool was much coarser. However, they had one great advantage: they grew and matured much faster. When the tide of sheep swept north of the Great Glen, to this was added the Cheviot, with its better fleece. Over the years both these Lowland breeds would develop distinctive Highland and Island strains.

The word *croft* is Lowland, meaning a smallholding. The first clearances were not altogether *as an fhearann* – off the land – but from the more fertile inland straths to small new holdings on the coast. It was a rough passage for the population, but without potato and lazy-bed to break in and clean the ground, it would have been impossible. People had their own plots of in-by or what passed for arable land, and still shared the grazing. The plots were deliberately kept small, so that people would be compelled to diversify their livelihood, by fishing, various trades, and not least, working at the kelp.

The French wars, first against the revolutionary Republic, and then against Napoleon, had created the circumstances for the Clearances. Coarse wool was needed for uniforms. Salt meat was

Clipping at Bourbladh, Morar, about 1910. The sheep are on sheering stools of stone and turf, an imported Borders fashion. Note also the more recent import of corrugated iron. M E M Donaldson, SEA.

needed to victual the fleet. The Spanish barilla which supplied the soda used to make soap could not be got, and *kelp* or seaweed supplied an alternative. It had to be cut at low tide, carried up onto the shore, dried and burnt. Evicted people were encouraged to settle near the kelping shores. It was very hard work, but afforded the opportunity for earlier marriage, because young people could set up house free of the older social restraints, and the population continued to expand.

With the end of the French wars in 1815, and the provision of a cheaper substitute for kelp by the Lowland chemical industry, demand for kelp dwindled. Now eviction took on a new and even more drastic character, not just to new crofts, but off the land altogether. Now the North-West of Scotland headed for the abyss. In 1845 the same potato blight which brought famine in Ireland came to Scotland. There was still sufficient unity of national feeling in Scotland for the Lowlands to organize relief and stave off the worst of the famine. But once the main emergency was over, a fresh round of evictions set in. Some heads of families tried to sustain the local populations, and went bankrupt. Tenant and laird alike were simply overtaken by events.

Some great speeches exist more in the minds of historians then they ever did in reality, and perhaps this is one. When the Duke of Sutherland came in person to recruit men for the Crimean War, he explained to a curious audience how the power of the Czar was growing dangerously, and ought to be reduced. Sitting behind a table laden with cash to be paid out in bounties, and with his clerks on either side, he waited for recruits to step forward. When they did not come, he grew anxious and then, becoming annoyed, he demanded an explanation. An old man got up and approached him. Among other things he explained that 'should the Czar of Russia take possession of Dunrobin and Stafford House at the next term, we could not expect worse treatment at his hands than we have experienced at the hands of your family for the last fifty years'. The few men that were left 'among the rubbish and ruins of the county has more sense than to be decoyed by chaff to the field of slaughter'. Defeated by this eloquence, the Duke rose, put on his hat and left the hall, leaving the three clerks to clear away the bounty money.

This incident was part of a rising tide of articulate protest that would lead eventually to the establishment of the Napier Commission which would recommend a vast reduction in rents, cancellation of arrears, and security of tenure. This became law in 1886. However, the Crofters Act did little for the dispossessed squatters who still clung illegally to the land in many of the island crofting areas. The vague promises of resettlement would count for little until after World War I. Yet the Act appears to have brought a badly needed element of stability. In Lewis, the most congested of the islands, the population began to drop for the first time, but in the empty straths of the mainland Highlands, there was now little population to return to whatever land might have been made available.

Paradoxically, the old man who had so bravely faced the duke was proved wrong. The people of the Highlands and Islands were loyal to the Empire. The Highland regiments would figure large in the British Army. Theirs would be among the bayonets that crushed the Indian Mutiny and supported the vast land-grabs of Africa.

After the middle of the century the sheep were becoming a less profitable enterprise. Overstocking had taken its toll, and as the agriculture of Australia and New Zealand developed, so did their exports of wool. The short-cut of the Suez canal opened in 1870, and faster and longer-range steamships and refrigerated transport meant that mutton was also in competition. The other side of the coin was Queen Victoria's and Prince Albert's love affair with the Highlands, and the interest of English sportsmen from earlier in the century. When the railways came to the Highlands, effecting direct communication with London, the stage was set for shooting lodge and deer forest, which despite the name was often treeless. The ancient forests such as Mar and Atholl were revitalized, and in other places the sheep were now evicted with shepherd and under-shepherd taken on as stalker and ghillie. It was a creation fuelled by romance. Some of the lodges, such as Ardverikie, were extravagant confections of Scottish baronial fantasy. Others such as Fealar at the back of Atholl were bleak, utilitarian, and very remote, reached over roads that wound for miles through now empty glens. That in a perverse way was now part of their attraction. Hours spent crawling over wet ground in total vigilance demanded stamina, and a poor shot at the end of it was a disgrace. The sport of kings was not an easy option. It also appeared to take place on another planet. Language, culture, interests, they were all different. It was part of the strange social schizophrenia that still marks Scotland today.

7 Home Fires.

With the end of the American Civil War in 1865 disbanded soldiers and land-hungry peasants from the old world headed over the coastal mountains into that vast space known as the Midwest to become homesteaders. The combination of railway, rifle and plough steadily emptied the prairie of buffalo and native Americans alike. By the 1870s, the railway yards of Chicago were starting to boom with the flow of farm produce eastwards.

To the American Midwest was added the pampas of South America, the wheat-lands of Southern Australia, and the lands of Manitoba, Saskatchewan and Alberta in Canada, and the South Island of New Zealand, the latter two countries the particular destination of many Scottish emigrants. The effect on European farming was drastic. Germany and France put up the shutters of protection. Britain stuck to free trade, because on balance then she still had more to gain than lose, nor could the produce of the Empire conceivably be excluded. There were also darker considerations. From the 1870s the British industrial cities were expanding at a phenomenal rate, and for the first time the urban population predominated. Exaggerated fears of the increasingly articulate industrial workers produced the appeasement of a cheap food policy. Imported grain and later cheap frozen meat kept coming. To the farming interest it was 'the Great Betrayal'.

To the farmers' woes were added bad seasons. For farmers the storm that caused the Tay Bridge disaster was remembered as the start of depression. Over the next few years a tenth of the best land went out of production. Yet by comparison Scotland was getting off lightly. The years of high investment, the advanced levels of agricultural engineering, the skill of the workforce, all made for what was probably still the most efficient farming in the world. The urbanized Central Belt offered a market for fresh dairy produce of the South-West, and no imported Argentinean beef could compete with the Aberdeen Angus for quality. The varied internal balance and the long rotations

An early wire-tying binder at Broughton Knowe, Biggar, probably 1900s. SEA

that had become characteristic of Lowland farming were a buffer against adversity.

Some of the developments that made American production cheap also applied to Europe. The self-knotting reaper-binder threw the sheaves off ready bound, saving labour. Mobile high-speed threshing mills became a familiar sight on the country roads, the whole *yoke* headed by a steam traction engine, with a baler as well as the mill, and at the back, the millmens' van where they ate and slept. Farm buildings underwent another phase of adaptation. The open cattle courts were often roofed over, and because labour was now accounted as a serious cost, the handling arrangements in the steadings became more streamlined.

Such was the pace of urbanization and change, that by the turn of the century, Great Britain produced only a quarter of her food. Far from going unnoticed, there were parliamentary enquiries into the security of imports in the event of war. Such was the faith and pride in the Royal Navy, the conclusion was that the supplies

A travelling threshing mill in Lauderdale. SEA

would get through. The German Emperor Wilhelm had ignored
Bismarck's advice to stick to Continental adventures and had
built a powerful navy that was not only a sure step towards con-
flict, but dictated one way in which it would be played out.

When the tragedy of the Great War began to unfold in August
1914, it had an unexpectedly direct effect on faraway places. The
British army was small, but in the years before the war it had
become much more professional. The full-time elements were
backed up by a well-trained volunteer reserve. This reserve was
particularly active in the Crofting Counties, because the annual
bounty was a useful contribution towards the rent. The activities
were also an attraction. The contests in marksmanship and the
annual camps were a break in routine and even a cheerful social
event. In Lewis and the Northern Isles, membership of the Naval
Reserve was also strong. Reserve members were not bound to serve
overseas, but most volunteereed to do so. The departure of the
reservists in late 1914 was dramatic. 'The men trickled in at most of

the stations in Caithness, and they became more numerous as the train entered Sutherlandshire. As the short winter day closed in, snow began to fall, and as the train wound through the valleys, all the houses were lit up, and the people stood at the doors waving and chanting.'

After their initial check, the German armies turned towards the Channel ports. The small British army which successfully barred their road was almost annihilated and the reserves which were rushed up contained many men who only a few weeks before had been bringing in the Hebridean harvest. The drive to recruit more men was highly successful throughout the countryside. In the end, two-fifths of Scotland's countrymen went to war, well over half the male population in the prime of life. There are places such as Stroma in Caithness or Benquhat in Ayrshire, where war memorials stand alone in an uninhabited landscape.

In late 1916 German submarines attacked all British-bound shipping to break the deadlock. The harvest of 1916 had been a bad one, with yields less than half those of 1914. In the spring of 1917, Britain faced starvation. Everything else was going wrong. The new volunteer armies were in tatters, and they were starved of reliable supplies. In this crisis Britain came to grips with reality, and took drastic steps to increase production of both food and war materials.

In the countryside indiscriminate recruitment was stopped, minimum wages guaranteed, and women's wages made equal to men's, to stop the drift to the munitions works in the towns. Retired men were got back to work, and boys were allowed to leave school before fourteen to work on the farms, and the Womens Land Army created. By these means the shortfall in the labour force dropped to 15%. The Corn Production Act guaranteed farmers minimum prices, and the hitherto somewhat inactive Agricultural Executive Committees got to work in every locality, coaxing the maximum production from every available scrap of land. Ground that had lain in grass for decades was ploughed up in a massive sowing programme in the spring of 1917. In the towns there was a swing from meat consumption back to the traditional diet of meal and tatties, thus saving cattle for the badly reduced dairy production. Grain production was up 20%, and in England where there was more slack to take up, it rose by a massive 40%. The potato harvest was a good one. To cap that, by the end of the year the U-boats had been beaten and the siege broken.

Here was a total reversal of the former muddle, and with that one curious yet significant victory. In 1918, the German army launched one last ferocious campaign. As the British reeled back and the storm-troopers overran the supply depôts crammed with food, their resolve crumbled. How could their starving country beat an enemy as well supplied as this?

The tragedy of the war lay not just in the loss of life. In the years before the outbreak, there had been signs of better things to

A volunteer in the Womens Land Army during World War I.
SEA

Wartime harvest in Orkney. Often old men, women and children were left to run the farms. The soldier is probably home on leave, hence the photograph. SEA

come. The competition between the new world and the old was reaching a balance, and the damage had all but run its course. The creation of a Scottish Board of Agriculture and the farm servants' and farmers' unions promised a more interested government policy towards farming, and a sensible framework for improving wages and housing.

After the war the framework remained but the sanity fled. By 1921 minimum prices and wages were abandoned, and the countryside was plunged into fresh depression. Now followed the biggest upheaval in land ownership since the Reformation. Until now the estate had been the main economic unit of the countryside, and most farmers were tenants. In the pre-war depression, many estates had become burdened with debt. The officer class had suffered the highest casualties during the war, and the loss was not just human but financial, in the form of death duties. For many old estates, it was the end of the road. Many sold out to the sitting tenants, who if they had been wise with the guaranteed income from the war years, were in a position to buy their land. Within a few years nearly two-fifths of Scotland had changed hands.

For the new owner-occupiers and remaining tenants alike, life remained a severe struggle. Towards the nadir of 1930, the prices of crops sometimes hardly repaid the effort of sowing them. Other farmers emigrated, not to the colonies, but the east and south of England, where farms on rich ground were going begging. Here the combination of acumen, professional skill and a hard-bitten work ethic triumphed, and the Scottish farmers prospered where their unfortunate predecessors had failed.

A colonization of a different kind happened on a modest scale in the Lowlands. There had first been moves in 1912 towards creating more smallholdings, but the landed interest had resisted this, fearing the 'crofterization' of Lowland Scotland. Now there was a more generous attitude among the lairds towards the men they had fought beside. The Lowland holdings were never economically significant, but the distinctive colonies of bungalows and barns became a familiar part of the countryside, and offered an independent if modest livelihood to many families.

For the crofting counties, this was a time of bitter paradox. Through the same legislation the people of the Hebrides in particular now looked at the prospect of repossessing the places from which their people had been evicted. The pressure was kept up through land raids or direct occupation of the ground. Lord Leverhulme's plans for the economic regeneration of Lewis and Harris foundered on this age-old attachment to the land. Then the depression closed in. In 1923 and '24 the *Metagama*, *Canada* and *Marloch* sailed for Canada laden with emigrants. All this was on top of the *Iolaire* disaster, when a shipload of returning veterans was wrecked within sight of home with a terrible loss of life. The names of these ships still signify heartbreak in the Hebridean memory.

There had been some attempt to bring in powered mechanization during the war, but the concept still had a long way to go before it could rival the horse, and the effect was negligible. What effort there was went into repair: the Forestry Commission was set up to make good the devastation of the war, and there was a drive to impose a discipline of hygiene on milk production which was a major source of infection. There were efforts to root out the widespread disease in the potato stock, and several important institutes of agricultural science were founded at this time.

Housing did not always improve and sometimes got worse. When tenants bought their land, they could not always afford repairs. Nevertheless, improving rural housing became a more pressing issue. Model plans were published, and the government smallholdings were coveted for their new houses as much as the ground, but as yet there was no legal compulsion to provide minimum standards.

Government initiatives to encourage voluntary activities are seldom rewarded with sparkling success. The Scottish Women's Rural Institute was a spectacular exception. As the motto 'For Home and Country' suggests, the movement was started during the war, with the intention of maintaining morale on the home front. It spread with great rapidity from East Lothian to the rest of the country, and was copied in England and Wales. Its popularity lay in the opportunity for women to have a social activity outside of the home, often for the first time ever. It was an organization which women ran themselves for themselves, and to begin with there was a lot of male sniping and opposition.

Even if the activities of the 'Rural' usually focused on domestic topics, that reflected the fact that entertainment was often just an extension of work. The songs sung at the *Kirns* or *Meal and Ale* celebrations and dances, or in the bothies, were set in the context of work. Ploughing matches were a celebration of work. Absorbing interests such as beekeeping were still an extension of food-production. A Sunday afternoon's walk would turn out to be an inspection of neighbouring farms, the regularity of their *dreels* and the neatness of the stacks in their cornyards. When stacks were built near a public road, special attention was paid to the quality of the rope-work, and the straw at the *easin* or bottom edge of the sloping part would be trimmed to a perfect fringe with sheep-shears against the critical eye of the Sunday strollers. The other entertainment was the *coortin*. That and work encompassed most people's lives.

Now the working day was generally limited to ten hours between leaving and returning to the stable, and a half day on Saturday was beginning to creep in. The annual week's *fair* enjoyed by the townsfolk was still an unthinkable luxury. New Year was a holiday, Sunday was the day of rest, and that with the odd holiday for a show or a ploughing match was it.

The rural population remained fairly stable in numbers, but young people were leaving for the towns. Even for the lively bothy men, a better life lay elsewhere, driving the now common buses and lorries, or getting a job in the *polis* in the towns, where the big steady country lads, the *meal mountains*, were considered good recruits.

Life was not just hard work, but was often attended with the worry of making ends meet, and even the stark reality of an empty cupboard. Yet living memory draws a picture of a positive life. Speech, manners and habits might appear *roch* to the modern eye, but honesty appears to have been the norm, as were the jokes and pranks that lightened the constant burden of work.

8 The Wee Grey Fergie

Tractors had long been on the go, ever since the invention of the internal combustion engine. At the turn of the century Britain was the biggest producer. The Great War had produced the 'Fordson', the first machine to *look* like a modern tractor rather than a sawn-off steam engine. The engine block and gearbox doubled as a chassis. In 1919 Wallace produced the 'Glasgow', designed by James Guthrie, which with its power drive to all three

A Fordson tractor at Overside of Fergus, Aberdeenshire. SEA

wheels was one of the most advanced designs of the day. Yet the tractor was still no more than an expensive iron horse. For the most part it worked with adapted horse implements. There was little or no advantage in it. Only in Buchan, scourged with endemic horse-sickness, did the tractor make much headway, giving that area the distinction of being the most highly mechanized part of Britain between the wars.

Combine harvesters also existed, and had been tried in the United States since the nineteenth century. There they were travelling mills with a cutter-bar, reel and canvas attached, pulled by monster teams of horses, and this provided the pattern for the first powered British combines made by Clayton and Shuttleworth in 1928. One came to Whittinghame Mains in East Lothian in 1932, where it worked well for many years, and this original is now in the NMS collection, one of the oldest combines in working order in the world. The combine offered tremendous advantages. The whole process of stooking, *leading* or bringing in the sheaves and building them into thatched and roped stacks, then pulling them apart again to thresh them: all this could be condensed into one operation. But the capital required was considerable, for not just the machine, but a grain-dryer. These tractor-hauled combines could only work at a slow rate, and the crop was liable to ripen ahead of them. Like the tractor, as yet they made little impression. Between the wars powered mechanization was still a realm of curiosity and experiment.

One experimenter was Harry Ferguson, born in 1884, the fourth son of James and Mary Ferguson, who farmed at Growell, Dromore in County Down. He served an apprenticeship at his older brother Joe's car and cycle repair workshop in Belfast in 1901. Ferguson had a maverick originality that amounted to inventive genius. He was interested in cars, aeroplanes – and tractors. In

The three-wheeled 'Glasgow', seen here being demonstrated in 1919, was the only tractor to be designed and built in Scotland. SEA

1933 he produced his prototype. Instead of being trailed as if by a horse, the implements were now mounted with hydraulic controls, so that they and the tractor became one machine. Implements could be raised and lowered at the touch of a lever, coupled, uncoupled, changed and moved by one man. For the first time, a whole system of implements could be designed to take advantage of the tractor's power. Ferguson formed a partnership with the Huddersfield engineering firm of David Brown, and the first Ferguson-Brown tractor was produced in 1936. Only 1,250 of these machines were made before Ferguson fell out with David Brown. He managed to interest Henry Ford in 1938, but no sooner had production got under way than another war broke out.

The same day that war was declared against Germany, the Secretary of State's signature set in motion a meticulously conceived plan of wartime agricultural production. Backed by a hierarchy of committees armed with considerable powers of compulsion, it got things done. Recruitment to the forces was carefully controlled and the labour force nurtured. The Womens Land Army was resurrected, and after a false start it swelled to over 8,000 land girls and 1,500 in the Timber Corps. The lessons of the Great War were reinforced by the guiding hand of scientific knowledge through the colleges of agriculture.

There was an invasion of the countryside not envisaged by the Nazis. The Poles who had escaped to renew their resistance against Nazi and Soviet occupation of their country, besides fortifying the east coast against invasion, did much to bring in the harvest in the critical years of 1940 and 1941. Many stayed after the war and made Scotland their home. From 1942 came the first Italian prisoners from North Africa, then Germans. They returned civility and often kindness with hard work. When peace came, there were 19,000 POWs at work on Scottish farms.

Building a hay ruck at West Windygoul, Tranent, 1923. The hay is lifted onto the top of the ruck by a pulley attached to the horse. SEA

Womens Land Army workers in East Lothian, World War II. Those in the middle are wearing uglies *or cane-framed bonnets to keep off the dust.* SEA

The mainstay of food production remained a man tramping behind a pair of horse, but powered mechanization was now more than a curiosity. Through the Lend-Lease arrangement with the United States tractors and combines as well as war materials came to Scotland. Some were allocated by the *Warags* as the executive committees were now known, and others went to the Government Tractor Service, flying squads that stepped in to boost the ploughing programme where it was falling behind. The short-term effect helped Britain survive the siege of the Nazi U-boats. The long-term effect was to build up a network of dealers, service and mechanical know-how.

In 1932 Walter Elliot had been appointed Minister of Agriculture. Himself an able farmer from Harwood, Bonchester Bridge, near Hawick, he had a full grasp of the desperate condition of farming. He knew of the numerous mass meetings that had united lairds, tenants and workers, and took immediate steps to restrict the dumping of imported agricultural produce. Under his stewardship the market for home produce became organized with legal backing. Liming and drainage schemes were introduced to retrieve the land from half a century of neglect. In 1937 the Agricultural Wages Board was set up, which specified not only rates of pay but also holidays. Things had come a long way since the comic uproar over the 'Turra coo'. In this incident in 1913 the farm servants as well as Robert Paterson, who farmed at Lendrum in Aberdeenshire, had refused to be a part of Lloyd George's National Insurance scheme.

When war came again, the worst of the depression had lifted. Faced with another massive effort of production, there was general agreement that there would no going back after the emergency was over, and the General Election of 1945 voted in a Labour Government. In 1947 an annual price review offered

A German POW carting potatoes at Devon, Kennoway, Fife . SEA

farmers a reasonable living. Since the 1870s, governments of all complexions had pursued a cheap food policy. Now for the first time they were willing to pay for it. The security created the framework for drastic change.

Harry Ferguson's agreement with the elder Henry Ford was based on a handshake. His grandson Henry Ford II had little time for gentlemen's agreements and, growing impatient with Ferguson, started to produce Ferguson system tractors on his own. But Ferguson's patents were secure, and he started his own production in the United States, made a deal with the Standard Motor Company in Coventry to start production in Europe, and secured £3 million from Ford in compensation. In 1946 the first TE20, the famous 'Wee Grey Fergie', came off the production line. Over the next ten years Standard made over half a million of them. Coupled with this came the first self-propelled combine harvesters. The Agricultural Revolution had matured into modern agriculture.

The change was disconcertingly rapid. By 1951, in the main arable areas of the Lowlands, horse-working was fast giving way to tractor and combine. By the end of that decade horse-working was unusual. Ten years on and it was a rarity to be photographed. With the horses went the horsemen and their families, and the various trades that had supported the old ways of working. The more adaptable smiths and mill-wrights changed into agricultural engineers or set up in the motor trade, but for others such as saddlers it was the end of the road. The old steadings that were built round the regime of horse-work would no longer be maintained by the traditional tradesmen, but would be replaced over the years with large pre-fabricated sheds with easy access for tractor and bulk-handling machinery.

The social consequences were far-reaching – in a word, depopulation. The difference in this latter-day clearance is that it happened in the context of post-war prosperity, and there were jobs in the towns to go to, and a building programme to provide housing. For those remaining on the land, this was double-edged.

Schools and village shops closed, and public transport declined, often with a quite drastic loss of community. But the jobs that remained changed in character and became much more long-term and secure, with a minimum wage set by the Agricultural Wages Board. Before the war, the six and twelve month fee had started giving way to the weekly wage, but now it went. The flitting from one farm to another at the term gave way to more settled habits, with a benefit to the education of farm workers' children. There were now substantial improvements in rural housing, and people could enjoy such things as bathrooms as a right and not a rare luxury.

The people who work the land in Scotland and the rest of the United Kingdom are now a minority even in the countryside. This is not so in other parts of the European Community. The founders of the Community made a decision to support rural farming populations as much as production. The intention was to maintain the stable fabric of rural life in unstable times, and to prevent an influx into urban areas struggling to rise from the wreckage of war. Neither policy ever reckoned with the extraordinary capacity of industrialized farming. The same acreage with only a fraction of the workforce can produce two and sometimes three times as much of a crop as it did nearly a century ago. With large areas of Western Europe farming this way, the surpluses of recent years are hardly surprising. Beyond that lies the continuing production of America and Australasia, and on top of that the extraordinary potential of Central and Eastern Europe.

Plenty has not brought lasting prosperity to all of those who have produced it. As the European Community struggles to contain production and control its agricultural budget, the shadow of depression is there. The grim reaper is debt incurred against plant and machinery that may not repay the investment, but without which production and income falls. High inflation, which brings with it high interest rates, is lethal for the farming community.

Although that community is now only a shadow of what it once was, paradoxically it is more tied to the land now than it ever was. Before World War I, less than a tenth of Scotland's farms were worked by owner-occupiers. Now nearly two-thirds of farms are owner-occupied. Where before the size of the labour force made many farms seem like small villages, now they are often the isolated homes of small families, and the work itself can be lonely. When the financial going gets tough, the image of the self-reliant countryman can wear thin, to reveal a vulnerable individual.

Yet if plenty brings its problems, the unpleasantness of shortage would soon concentrate the mind, and the horrors of famine destroy it. As with other industry, modern agriculture runs on oil: everything from oil for tractor and combine and forage harvesters to fertilizers and the bags they come in. In the future there may well be things we will have to discover from the past.

FURTHER READING

ALEXANDER, W *Johnnie Gibb of Gushetneuk*, Douglas, 1880

COPPOCK, J T *An Agricultural Atlas of Scotland*, Edinburgh 1976

DEVINE, T M ed *Farm Servants and Labour in Lowland Scotland*, Edinburgh 1984

FENTON, A *Country Life in Scotland*, Edinburgh 1987

FENTON, A *The Island Blackhouse*, Edinburgh 1978

FENTON, A *The Northern Isles: Orkney and Shetland*, Edinburgh 1978

FENTON, A *The Shape of the Past*, 2 vols, Edinburgh 1986

FENTON, A and B WALKER *The Rural Architecture of Scotland*, Edinburgh 1981

GRANT, I F *Everyday Life on an Old Highland Farm*, revised edition, London 1981

GRANT, I F *Highland Folk Ways*, London 1975

HUNTER, J *The Making of the Crofting Community*, Edinburgh 1976

MARTEN, B *Harry Ferguson*, Ulster Folk and Transport Museum, nd

MARTIN, A *Kintyre Country Life*, Edinburgh, 1987

MORTON, R S *Traditional Farm Architecture*, Edinburgh 1976

NAISMITH, R J *Buildings of the Scottish Countryside*, London 1985

ORR, W *Deer Forests, Landlords and Crofters*, Edinburgh 1982

PARRY, M L and T R SLATER, eds *The Making of the Scottish Countryside*, London and Montreal 1980

POWELL, B *Scottish Agricultural Implements*, Princes Risborough, nd

SANDERSON, M H B *Scottish Rural Society in the Sixteenth Century*, Edinburgh 1982

SMOUT, T C *A Century of the Scottish People, 1830-1950*, London 1986

SPROTT, G *Robert Burns, Farmer*, Edinburgh 1990

SYMON, J A *Scottish Farming*, Edinburgh 1959

WHYTE, I *Agriculture and Society in Seventeenth-Century Scotland*, Edinburgh 1979

PLACES TO VISIT

The following museums and heritage centres display material related to farming and country life in Scotland.

Aden Country Park, Aberdeenshire: *North East of Scotland Heritage Centre*

Bettyhill, Sutherland: *Strathnaver Museum*

Biggar, Lanarkshire: *Gladstone Court Museum*

Blair Atholl, Perthshire: *Atholl Country Collection*

Brodick, Isle of Arran: *Isle of Arran Heritage Museum*

Ceres, Fife: *Fife Folk Museum*

Cousland, Midlothian: *Cousland Smiddy*

Dunbeath, Caithness: *Lhaidhay Croft Museum*

Edinburgh: *Scottish Ethnological Archive, NMS*

Fort William: *West Highland Museum*

Gairloch, Ross-shire: *Gairloch Museum*

Glamis, Angus: *Angus Folk Museum, National Trust for Scotland*

Glencoe village, Inverness-shire: *Glencoe and North Lorn Folk Museum*

Glenesk, Angus: *Glenesk Museum*

Harray, Orkney: *Corrigal Farm Museum*

Ingliston, Edinburgh: *Scottish Agricultural Museum, NMS*

Inveraray, Argyll: *Auchindrain Museum of Country Life*

Crofter's pony and cart at Ness, Lewis, Ross-shire, 1920s or 30s. The sides can be taken off to make the cart flat. SEA

Keensgarth, Shetland: *Tingwall Agricultural Museum*

Kingussie, Inverness-shire: *Highland Folk Museum*

Port Charlotte, *Islay: Museum of Islay Life*

Port of Ness, Isle of Lewis: *Ness Historical Museum*

Portree, Skye: *Skye Cottage Museum*

Shawbost, Isle of Lewis: *Shawbost School*

Stornoway, Isle of Lewis: *Museum nan Eilean*

Tain, Ross-shire: *Tain Museum*

Thurso, Caithness: *Heritage Museum*

Voe, Shetland: *Shetland Croft House Museum*

In addition to the NMS material in the *Scottish Agricultural Museum* and the *Scottish Ethnological Archive*, York Buildings, Queen Street, Edinburgh, there will be displays relating to farming history and rural life in the new *Museum of Scotland* due to open in 1998. The NMS is currently investigating possibilities for a working farm, which would show farming and country life in the 1950s.